PERMEABILITY AND ELECTRIC PHENOMENA IN MEMBRANES

BY

LEONOR MICHAELIS

PROFESSOR OF BIOLOGICAL CHEMISTRY, UNIVERSITY OF BERLIN,
AND PROFESSOR OF RESEARCH MEDICINE, JOHNS HOPKINS UNIVERSITY

Lecture delivered at Columbia University in the special course on Contemporary Developments in Chemistry given in the Summer Session of 1926 on the occasion of the opening of the Chandler Chemical Laboratories

COLUMBIA UNIVERSITY PRESS

NEW YORK

1927

Copyright, 1927
By COLUMBIA UNIVERSITY PRESS

Printed from type. Published March, 1927

Printed in the United States of America

PERMEABILITY AND ELECTRIC PHENOMENA IN MEMBRANES

by

LEONOR MICHAELIS

1. Sieve and Homogenous Membranes

For the purpose of this lecture, let us define a membrane as a thin layer of a substance of a certain solidity separating two liquids from each other, and of such quality as to prevent the two liquids from mixing when shaken or stirred, or in other words, such as to prevent *convection,* while allowing diffusion of some or even all kinds of the molecules contained in the solutions. When only the one side of such a membrane is in contact with a liquid and the other with the air a membrane may function also as a filter. However, this case is not to be the proper subject of this lecture though closely related to it. The simplest way of realizing a membrane is the idea of a sieve the pores of which are narrow enough to prevent the convection. According to the size of the pores, larger or even smaller particles representing components of the liquids, are prevented from convection across the membrane. When the pores are narrow enough even some kinds of the dissolved molecules will not penetrate, and in this case it may happen that the membrane functions as an osmometer allowing the solvent to go through the membrane to such an extent that the difference of the levels compensates the osmotic pressure of the impermeable molecules. In fact, most of the membranes may be represented by sieves. However, there are membranes also of a homogeneous structure. Much has been said about the so-called lipoid membranes which are supposed to be a homogeneous membrane of a substance such as oil, wax, lecithine, etc., separating two aqueous solutions and being impermeable for any substance insoluble in the lipoid membrane substance. Though it must not be doubted that a membrane may work by this mechanism of selective solubility and though many instances have been shown by Overton and H. H. Meyer in membranes of living cells, yet the frequency of the occurrence of a membrane of this type without any sieve effect seems to me to be overestimated. Recently Kahlenberg described a selective dialysis through a membrane of stretched rubber dam. When different substances are dissolved in pyridine and separated from pure pyridine by a rubber membrane, only such substances are prevented from dialysing which are difficultly soluble in lipoids but easily soluble in water. This seems to be an example of a homogeneous membrane. However, it should be borne in mind that even in a rubber membrane the selective permeability of this kind occurs only in pyridine solutions. This membrane is not able to separate the easily water-soluble and the easily lipoid-soluble substances when these are dissolved in water. Later on I shall describe a kind of rubber membrane which quite certainly behaves like a sieve. Even in the pores of a sieve the so-called lipoid substances may sometimes behave differently from the water-soluble substances, because the lipoid substances in general are capillary active at the same time and have different properties with respect to the wetting of the walls of the pores. So, the sieve nature of certain membranes is quite obvious, while the homogeneous nature of a membrane very often cannot be definitely proven.

2. The Size of the Pores in Sieve Membranes

A pure sieve membrane exerts its effect by the fact that the substance of the membrane, itself, is practically impermeable to all substances concerned. It is only its holes or pores or channels which allow the penetration. Some examples of membranes in which the size of the pores differs are given here. A clay diaphragm prevents any convection but allows the diffusion of practically any dissolved substance, even of many colloids. It is used in electrochemical engineering as a means of preventing the mechanical mixing of the products of electrolysis without preventing the ions of the solution from following the forces of the electric field. Much narrower are the pores of parchment paper and of certain dried animal membranes such as gut, and pig's bladder, which are impermeable for colloids, or better, the impermeability through which seemed to Graham to be the characteristic of what he called colloids. Non-colloidal molecules permeate through these membranes and the presence of the membrane does not appreciably alter their rate of diffusion although, later on, we will discuss some smaller effects even in this regard. A third range in the size of pores is represented by such membranes as copper ferro-cyanide, discovered long ago by W. Traube and used by Pfeffer as an osmometer, which led van't Hoff to evolve the classic theory of osmotic pressure. This membrane is approximately impermeable to any dissolved substances and permeable for water alone. However, the discoverer himself and recently Collander in a new series of experiments found this membrane to be permeable for smaller molecules such as potassium chloride and urea. In the series of the aliphatic alcohols the permeability decreases with increase in the molecular weight. Larger molecules such as sugar are not at all able to penetrate this membrane. In a certain sense the following kind of membrane which I found recently may be placed in the same range. This is a membrane of collodion which in contrast to the well-known collodion membrane usually used for dialysis and ultrafiltration, has been completely dried so as to remove the last traces of ether or alcohol before it is brought in contact with water. Such membranes were considered as completely impermeable and useless, hitherto. Yet they turned out to be of quite a peculiar interest. In regard to the permeability to the different kinds of dissolved molecules this membrane behaves very similarly to the copper ferrocyanide membrane. Generally all diffusion through these membranes proceeds very slowly but still there are great differences in the permeability according to the size of the molecules. Table I shows the approximate relative diffusion coefficients of some substances, the first series represents data obtained with a very easily permeable collodion membrane of the old, well-known type. These coefficients may be supposed to be approximately equal to those of free diffusion without any membrane. The second series gives the relative diffusion coefficients in the dried collodion membrane. The order of the substances arranged according to their diffusibility in a very dilute aqueous solution is the same in both cases, but the differences between the substances are enormously increased in the case of the dried

TABLE I

	Methyl-alcohol	Ethyl-alcohol	Propyl-alcohol	Butyl-alcohol	Urea	α-Mono-Chlor-hydrine	Glucose
Relative diffusion coefficient for the ordinary easily permeable collodion membrane..........	1.22	1.15	1.0	1.0	1.0	0.7	0.5
Relative diffusion coefficient within a *dried* collodion membrane..	9.2	4.1	3.0	0.82	1.0	0.07	0.000

collodion membrane. Note the great differences between methyl, ethyl, and prophyl alcohols. This membrane is completely impermeable to molecules of the size of any sugar. The only most evident contrast to the copper ferrocyanide membrane consists in that the dried collodion membrane is extremely difficultly permeable even to water, which fact certainly has something to do with the fact that dried collodion is not easily wetted by water.

3. Electric Potential Differences of the Two Sides of a Membrane

A remarkable phenomenon in membranes which is not only interesting by itself but also leads to important conclusions, is the difference of the electric potential on the two sides of the membrane when in contact with two different solutions. In general, when two solutions of different composition are in contact even without any membrane, a potential difference is established, which fact plays an important part in the determination of the electromotive force of galvanic chains in which liquid junction potentials sometimes give rise to an inconvenient complication. When the two solutions are separated by a membrane, the potential difference is, as a rule, different from the one established in free contact. Let us restrict ourselves to the most simple cases, as such simple cases wholly suffice for recognizing the essential mechanism of the membrane in this respect. There are two such simple cases: either the two solutions contain only one and the same electrolyte but in different concentrations, or the two solutions each contain a different electrolyte but in equal concentration. When there are two solutions of the same electrolyte in different concentrations the potential difference is produced by the difference of the mobility of the cation and the anion. While diffusion is taking place the more mobile ion moves ahead, charging the more dilute solution with electricity of its own sign. The potential difference thus established amounts to

$$E = \frac{u-v}{u+v} 0.057 \log \frac{c_1}{c_2} \text{ volts,}$$

where c_1 and c_2 are the concentrations of the two solutions, and u and v are the mobility of the cation and of the anion. There are two limiting cases. When the mobilities u and v are equal, then the potential difference is zero. On the other hand, when the mobility of one of the two ions is relatively so small as to be negligible with respect to the other, the potential difference reaches its maximum value, amounting to $\pm$ 57 millivolts if the ratio of the two concentrations is 1:10 (the sign depends on whether the mobility of the cation or of the anion is negligible). Generally the effect of a membrane separating such solutions is to change the potential difference from that obtained when the two solutions are in contact without a membrane. That may be interpreted by admitting the ratio of the mobilities of the two ions to be different within the pores of the membrane and in a free solution. The simplest case is a chain with two solutions of KCl, 0.1 N and 0.01 N. As the mobilities of K^+ and Cl^- are equal, there is no potential difference in free contact. Almost every membrane interposed between these solutions produces a potential difference, mostly in the sense that the more dilute solution becomes positive. Hence we are led to the interpretation that the mobility of the Cl^- ions becomes relatively smaller than that of the K^+ ions. The absolute value of the mobility being of no importance for the potential difference, the conclusion only concerns the change of the mobilities of K^+ and Cl^- relatively to each other. There is sufficient reason for assuming that the *absolute* value of the mobility is decreased by the membrane even with the K^+ ion, but that the mobility of the Cl^- ion must be decreased still more. An ordinary clay diaphragm shows only traces of such an effect. The usual large-pored collodion membrane may produce a small potential difference of 5–10 millivolts, a parchment paper membrane one of 10–15 millivolts. When one prepares a membrane of solid paraffine by impregnating filter paper with molten paraffine, the paraffine when solidifying seems to shrink a little and to form very fine ultramicroscopical pores, and a real sieve membrane arises which is permeable even to sugar; from this we obviously recognize that the permeability of a substance does not have any connection with its solubility in paraffine. Such a membrane gives a potential difference of about 13 millivolts under the conditions mentioned before. A membrane of wax prepared in the same manner gives 21 millivolts; filter paper impregnated with mastic gives 25 millivolts, with India rubber about 30 millivolts, and finally a membrane of dried collodion gives 50–55 millivolts, thus almost reaching the theoretical maximum value of 57 millivolts which is reached when the membrane is entirely impermeable for the anions. Chains with other salts give results which always lead to the same conclusion that the anions become relatively less mobile than the cations when permeating the membrane.

The impermeability for any anion, of the dried collodion membrane appears to be the realization of an old idea of Ostwald, quoted many times by biologists in order to explain certain properties of the cell membrane. This impermeability ought to be proved by direct diffusion experiments. Such an experiment is more difficult than may be anticipated, because the speed of diffusion of even the most mobile ions is so small that with membranes of sufficient thickness to secure a certain mechanical solidity some weeks are necessary to get definitely determinable amounts across the membrane, and within this period a change of permeability of the membrane often takes place which can be recognized by the fact that the maximum effect of the potential difference in the chain described before is no longer obtained. If we were able to make membranes as thin as a cell membrane and yet sufficiently mechanically resistant, the diffusion experiment would be easy. But that unfortunately is not the case. However, such an experiment which is quite reproducible in a series of selected, good membranes can be effected. When a .1 N solution of hydrochloric acid is separated from pure water by such a membrane, no acid diffuses through the membrane even in 2–3 weeks. The anions do not diffuse and do not allow the H-ions to diffuse either on account of the electrostatic attraction. However, when a solution of hydrochloric acid is separated by the membrane from a solution of potassium chloride, the potassium chloride solution becomes acid, this acidification progresses slowly but markedly from day to day. Here the H^+ and the K^+ ions are exchanged through the membrane. There is no diffusion of the hydrochloric acid in the usual meaning of the word. I hope to be able to improve the technique of such diffusion experiments in general in order to support the assumption of the immobility of the anions.

The second kind of simple chains is obtained when the two solutions each contain a different salt but in equal concentration. For the sake of simplicity let us take two salts with a common ion; e.g., HCl and KCl, or KCl and K_2SO_4. In such a case the potential difference should be

$$E = \pm\, 0.057 \log \frac{u_1 + u_c}{u_2 + u_c},$$

where u_c means the mobility of the common ion of the two salts, and u_1 and u_2 the mobilities of the two different ions. According to whether the common ion is the cation or the anion, the positive or the negative sign in this formula is valid. Such chains, simple as they may appear, yield results less easily interpreted in general, as there are three different mobilities to be dealt with. The conditions may be easily understood only in the case of the dried collodion membrane, as the mobilities of the anions may be set $= 0$. Now, when a *cation* is the common ion of the mobility u_c, then $u_1 = u_2 = 0$, and hence

$$E = 0.057 \log \frac{0 + u_c}{0 + u_c} = 0.$$

This calculation is practically confirmed by experiment for any combination of salts with any kind of ions provided the cation is the common ion. In opposition to the case when an *anion* is the common ion and the cations are different and possess the velocities u_1 and u_2, the potential difference becomes

$$E = 0.057 \log \frac{u_1}{u_2} \text{ volts.}$$

So we may approximately compute the relative mobilities of the two cations within the membrane, from the measurements of the potential difference, of course only when using a membrane in which the mobility of the anion has actually proven to be negligible by the measurement of a chain of the first type. Table II shows the relative values of the mobility of the univalent alkali

TABLE II

	Li^+	Na^+	K^+	Rb^+	H^+	Ratio $Li^+ : H^+$
Relative mobility in a simple aqueous solution	0.52	0.65	1.00	1.04	4.9	1 : 9.4
Relative mobility within the dried collodion membrane	0.048	0.14	1.00	2.8	42.5	1 : 890

metal ions. The first line gives the values for the free diffusion as known for a long time by the determinations of transfer numbers. The second line shows the relative values of the mobility within the membrane. Obviously the order of increasing mobility is the same, but the differences in the mobilities of each two members of the series is largely increased in the membrane. The greatest difference is the one between Li^+ and H^+. The ratio of their mobilities in free diffusion is 1 : 10, but in the membrane almost 1 : 1000. Even the ratio of Na^+ and K^+, which in free diffusion is 1 : 1.5 reaches in the membrane about 1 : 7. Some dried collodion membranes have also been found in which the effects are somewhat smaller.

All of the kinds of membrane give the common effect of retarding the anions. When we consider all these membranes we become aware of the fact that all of them consist of a material which under different experimental conditions has proved to be electronegative against most of the aqueous solutions. But there are also electropositive membranes. It is very remarkable that we do not know of any membrane which is electropositive against an aqueous solution under all or at least most conditions but only of amphoteric membranes such as membranes of proteins, casein or coagulated albumin, alumina, animal membranes consisting of gelatine—or protein-like substances. Such membranes take a positive charge in acid solutions, when the hydrogen ion concentration is greater than the one of the isoelectric point of the ampholytic substance. It is a remarkable coincidence that the influence of these membranes on the diffusion potential when in contact with two different solutions is also shifted when the hydrogen ion concentration passes from the one side of the isoelectric point to the other. This has been pointed out by Höber and Mond for diaphragms of casein or globulin, by Fujita in my laboratory for gelatine and coagulated egg albumin. The effect of such membranes investigated by the method of measuring the diffusion potential is such as to lead to the conclusion that these membranes retard the anions when in contact with a solution on the alkaline side of the isoelectric point and retard the cations when on the acid side of it. Another example of the reversion of the properties in an amphoteric membrane was shown by Girard a long time ago. When barium chloride is separated by an animal membrane from pure water and the barium chloride solution is slightly acidified, the Cl^- diffuses much faster through the membrane than Ba^{++} does. But when the solution is slightly alkaline, the Ba^{++} migrates faster than Cl^- does.

These effects on the diffusion potential differences may be explained as follows. It has been known for a long time, since electroendosmosis was experimentally investigated by Quincke and theoretically explained by Helmholtz, that these membrane-forming substances take an electric charge when in contact with an aqueous solution, this charge generally being negative but in some cases also positive. Nowadays we represent the electric double layer at the boundary of two phases as two sets of ions. Take for instance collodion which is negatively charged against an aqueous solution. We presume that the collodion absorbs a layer of negative ions from the solution, hydroxyl ions or even other anions, and thus by electrostatic attraction, causes the positive ions of the solution to form another layer adjacent to the absorbed layer of anions. The first absorption layer of the anions is best thought of as a monomolecular layer of ions. Harkins and Langmuir's idea of a monomolecular layer of molecules at absorbing interfaces may justly be expanded likewise to absorbed *ions*. This layer may be considered to be strongly attached to the absorbing wall so as to be practically immovable against the wall. However the adjacent layer of positive ions is located in the fluid and movable part of the liquid. It is not a real plane but a layer of a certain thickness. It consists only in that the concentration of the cations in the liquid is higher on the side facing the boundary of the liquid than in the main part of the liquid. The channels or pores of the membranes possess the same distribution of ions in the form of a double layer. The anions of the solution inside the channel are partially fixed by the wall of the channel losing their mobility while the cations remain freely movable. Let us assume as a first approximation, that the mobilities of the movable ions within the capillary channel are the same as in a free solution, still those of the anions which are fixed by the wall will not participate in any diffusion. Thus the average mobility of the anions is decreased. This average mobility is the only magnitude which can actually be measured in terms of the potential differences described above. Therefore u and v in the above equations mean average mobilities. According to a general law of absorption the percentage rate of absorption of the ions will be the greater the more dilute the solution. When the channels are narrow enough and the electrolyte concentration of the liquid is small enough, it may be that practically all of the anions are absorbed and immovable. In this case we attain the maximum value of the diffusion potential, such as described before for the extremely fine pored dried collodion membrane, this effect being approximately 57 millivolts for a concentration range of 1 : 10. It is in best agreement with this theory that this maximum effect is most easily reached in dilute solution, when for instance we have a 0.01 N and a 0.001 N solution, while in the combination 1 N : 0.1 N the maximum effect even in very small pored membranes can never be reached.

We assumed in the beginning that the mobilities of those ions which are freely movable in central portions of the channels are the same as in a free solution. This assumption is, of course, only very rough and we have to correct it according to the results of experiments with a collodion membrane between two solutions of different electrolytes such as hydrochloric acid and potassium chloride in equal concentration. These experiments were interpreted by us on the assumption that the difference of the mobilities in the pores is greatly exaggerated as compared with the differences in free water. That may be conceived by applying our knowledge on the causes of the differences of mobilities in general. Schematizing a little we may imagine every ion dragging along a shell of water when moving through the solution, the diameter of the water shell dragged along varying with the chemical individuality of the ion. Among the alkali ions the water shell becomes greater according to the order Li^+, Na^+, K^+, Rb^+, Cs^+. The velocity of an ion is determined by the extent of the surface of the dragged water shell, this water shell being compelled to overcome the frictional resistance of the surrounding water. I should like to emphasize that this idea is a little schematized since there is no definite water shell sharply confined to the surrounding water molecules. However our manner of representation though not perfectly adequate is sufficient for the present purpose. Now, when the channels are wide enough to allow the cations themselves but not the surrounding water shell to enter, the frictional resistance will increase with the magnitude of the water shell which the ion drags along, or as the attractive force for water by the ion becomes greater. There is a layer of water fixed at the wall in any case, such as in any capillary which is wetted by water, and the attraction of this unmovable water layer for the moving ion represents an additional frictional force against the motion of the ion.

4. The Glass Membrane

A membrane very remarkable in its electromotive effect is a very thin membrane of common glass which can be easily made by blowing a glass tube in the usual manner of the glass blower. The conductivity of such a membrane is very poor. Potential differences can be measured in general only by electrostatic methods such as by the quadrant electrometer. Especially between two solutions of different hydrogen ion concentrations, high potential differences can be observed at the two sides of the glass. The first one to describe this effect was Cremer and particularly through the investigations of Haber and Clemensievitsch it has been pointed out that the potential difference is approximately the same as in a hydrogen ion concentration chain with gas electrodes. Haber's theory is as follows. Traces of water but not of the dissolved electrolytes penetrate the glass wall which may be considered as a colloid with a certain, though small faculty of swelling in water. So within the membranes hydrogen and hydroxyl ions are present as a consequence of the dissociation of the water, but these ions are in a quite constant concentration not depending on the concentration of the hydrogen ions in the outside solution. This condition being given a simple thermodynamical consideration shows that solutions of different hydrogen ion concentrations on the two sides of the glass should behave just as they do in a hydrogen gas chain. The perfect realization of the potential difference expected from this theory meets some difficulties and seems to depend on the kind of glass. However sometimes it may be realized in fact. Some authors even recommended the glass for pH measurements in physiological liquids. It has to be supposed, for this purpose, first that the theoretical maximum value of the potential difference is really obtained, secondly that no other ions except the hydrogen ions control the potential difference, particularly not those of sodium and calcium. On the other hand several authors described the effect as not quite reaching the expected value and pointed out that the potential difference depends on other cations also. The anions were without any influence at all. Thus it seems that sometimes the glass membrane may give effects rather similar to the dried collodion membrane. I should suggest, therefore, to apply the theory developed before for the collodion membrane, likewise for the glass membrane, as a preliminary trial which ought to be proven by further investigations. Even in the collodion membrane the hydrogen ions turned out to be by far the most permeable ions. If we assume the glass membrane to behave as a sieve membrane with still smaller pores than collodion it may happen, in the limiting case, that only hydrogen ions can penetrate the membrane. If that occurs the maximum effect such as obtained by Haber can also be explained by the theory developed for collodion, and these theories are not contradictory at all but only different representations of the same idea. However the theory which I suggested is more general, including the other as a special or limiting case.

5. Phenomena of Electric Polarization in Membranes

When a membrane is in contact with solutions of the same composition on either side no potential difference is set up and nothing happens, of course. When we close such a system into an electric circuit the current flowing across the membrane will in general produce some alterations in the composition of the solutions. In order to understand this phenomenon we may simply consider the membrane as a third phase interposed between the two solutions and assume the relative mobilities of the different ions to be different from the one in the aqueous phases. It is of no importance for this consideration whether the membrane is really a new homogeneous phase, or a channel system, and whether the considered mobility is but an average mobility.

Many years ago Nernst and Riesenfeld showed that a phase interposed between two solutions of the same composition must produce a difference in the electrolyte concentration on the two sides of the middle phase when an electric current flows, whenever the transfer numbers of the ions in question are different within the membrane and in the solutions. A special case of that is a phenomenon first described by Hittorf and carefully studied by Bethe and Toropoff, a change of the hydrogen ion concentrations. The one solution becomes acid, the other alkaline, in the neighborhood of the membrane. This effect plays an important rôle in the practice of electroendosmosis and of the electrodialysis recently recommended by W. Pauli for purifying colloids. When the membrane is electronegative the liquid of the cathode side of the membrane becomes acid, the one of the anode side becomes alkaline, opposite to the changes taking place at the two metal electrodes. The effect is greater the more dilute the solutions are and it can no longer be observed at all in stronger solutions, such as, let us say, 0.1 normal salt solutions. In membranes which are electropositive against the solution, the effect is reversed. When the solutions are such as to correspond to the isoelectric point of the membrane—supposing it is an amphoteric membrane—this effect vanishes completely. This alteration of the hydrogen ion concentration is a part of the general phenomenon of the change in concentration of any ion through the effect of a membrane. However it must necessarily be supposed that the concentration of the dissolved electrolytes is so small that the participation of the ions of the water itself in conducting the current is appreciable. For that reason this effect vanishes in electrolyte solutions of higher concentrations.

6. Anomalous Endosmosis

I may assume that the phenomenon of electroendosmosis through membranes is so well known, that a description of it is unnecessary here. It may suffice to state that endosmosis consists in the transport of water through a porous membrane when an electric current flows through the membrane interposed between any aqueous solutions. But another apparently related phenomenon may be mentioned a little more in detail. When a membrane separates a solution from pure water, a stream of water flows from the pure water into the solution. This arrangement can be used for measuring the osmotic pressure, if the dissolved substance cannot diffuse at all through the membrane. But also when the dissolved substance is permeable, such a stream of water can flow in the beginning of the diffusion process provided the permeability for water is greater than the one for the dissolved molecules, *e.g.*, when a common, well permeable collodion membrane separates pure water from a sugar solution, sugar diffuses into the pure water. But in the beginning of the process the diffusion of water into the sugar solution not only takes place at the same time but it persists as well and is the most obvious phenomenon. The cause of this phenomenon needs no further explanation. However sometimes it happens that this water flow has the opposite direction, *e.g.*, from the solution into the water, in opposition to what may be expected according to the osmotic pressure. This so-called anomalous osmosis occurs only in electrolyte solutions, never in such solutions as sugar. Several examples of anomalous osmosis have been on record for a long time, but this problem reached a stage of a better experimental treatment only a few years ago when Jacques Loeb described it for a usual, relatively large-pored collodion membrane which has previously been in contact with a protein solution such as gelatin. The pores of such a membrane are covered with an absorbed layer of the protein which cannot be removed by washing. Now, there are some solutions which show anomalous endosmosis in such a membrane, for instance a very dilute solution of hydrochloric acid or of a salt of a trivalent cation, that means a solution of such a kind as to charge the gelatinized membrane positively. This phenomenon can only be observed when the solutions are very dilute. But also, in such electrolyte solutions which do not cause a water stream in the direction opposite to the osmotic pressure, the osmosis is sometimes anomalous in such a membrane insofar as the speed of the flow of water is greater or smaller than the speed of the water current which is produced by the osmotic pressure of a non-electrolyte such as sugar of the same osmotic pressure. This effect, also, disappears when the concentrations are higher, then the usual osmotic effect takes place again. Loeb recognized that besides the usual osmotic effect, in electrolyte solutions there is another effect which is of electrical nature and interferes with the first one and according to the conditions may increase or decrease the velocity which would be produced by the osmotic pressure alone. A very careful study has been published on this subject by Bartell, who investigated pure, collodion membranes which were not coated with protein. The effect is smaller in this case, as a rule, not reversing the direction of the expected water stream, but only diminishing or increasing it. The whole phenomenon is very interesting and may ultimately explain many processes of resorption and secretion in living organisms, which processes often take place against the osmotic pressure. The explanation given by Loeb and by Bartell consists in assuming an electroendosmosis interfering with the common osmosis. Now, electroendosmosis requires the existence of a steady electric current through the membrane. There is a potential difference across the membrane, as we have shown before. But which conducting material closes this potential difference to a real electric current? Freundlich and Nathanson assume that the substance of the membrane may function as the conductor. This explanation does not hold at all for the collodion membrane. For collodion owes its conductivity only to the water content of the pores. Dry collodion happens to be an ideal insulator, I dare say a better one than even paraffine. Now, an electric current in a liquid medium means the movement of cations in one direction without an equivalent stream of another kind of cations moving in the opposite direction, or without the moving cations dragging along the equivalent amount of anions in the same direction. How can such a process take place in a membrane simply separating two liquids? Speaking candidly, we do not know yet. There is an explanation given by Bartell. He assumes, the outer layer of the double layer at the wall of the pores which is freely movable in the liquid, to be the conductor. The electric current is supposed to be a circuit moving along the axis of the cylindric channel in one direction and along the wall of the cylinder in the opposite direction. We may be certain that there is something right in this hypothesis, but we cannot understand it completely. Perhaps this lecture may be an incentive for some of you to ponder a little on that problem and to attempt to solve it yourselves.

Bei Fragen zur Produktsicherheit wenden Sie sich bitte an:
If you have any questions regarding product safety,
please contact:

Walter de Gruyter GmbH
Genthiner Straße 13
10785 Berlin
productsafety@degruyterbrill.com